ASTRONOMY AND COSMOLOGY

Book 3

THE TEACHING OF DJWHAL KHUL

* * * * *

The Teaching of the Master of the Trans-Himalayan Esoteric School, Djwhal Khul. The continuation of books of HP Blavatsky and AA Bailey. Synthesis of Science and Religion

* * * * *

TATIANA DANINA

* * * * *

CREATE SPACE

* * * * *

The Teaching of Djwhal Khul – Astronomy and cosmology
Copyright © 2014

Translation from Russian by Tatiana Danina

The contact information

https://authorcentral.amazon.com/gp/books?ie=UTF8&pn=irid58388648

danina.t@yandex.ru - e-mail

Facebook: https://www.facebook.com/tatiana.danina

Vkontakte: https://vk.com/t.danina

The books of the series "The Teaching of Djwhal Khul – Esoteric Natural Science" - **"The main occult laws and concepts"** - http://www.amazon.com/Main-Occult-Laws-Concepts-ebook/dp/B00GUJJR72

(paperback - http://www.amazon.com/The-Teaching-Djwhal-Khul-concepts/dp/1499625421

"Ethereal mechanics" - http://www.amazon.com/The-Doctrine-Djwhal-Khul-mechanics-ebook/dp/B00I8KSY8Y (paperback - https://www.createspace.com/4836813)

"New Esoteric Astrology, 1" - http://www.amazon.com/dp/B00JF6RMCY (paperback - https://www.createspace.com/4827294)

"Thermodynamics" - http://www.amazon.com/dp/B00KGHK8EU (paperback - https://www.createspace.com/4838412)

And here is the book of my grandpa, **Michael Novikov, a military paramedic**. You can read his **memories about the Finnish war** http://www.amazon.com/dp/B00JYDITQ6

(paperback - http://www.amazon.com/Memories-Russian-Military-Paramedic-Michael/dp/1499786115)

We wish you enthralling reading!

TABLE OF CONTENTS

01. Dispute between theories of heating and cooling the planet.

02. Nuclei of Galaxies and stars – a comparative characteristic.

03. The mechanism of rotation of planets.

04. On what does the heating of planets by stars, of stars by galactic nuclei and of nuclei of galaxies by nuclei of supergalaxies affect?

05. Reasons for beginning of the rotation of planets.

06. Retrograde and prograde rotation of planets.

07. The seasonal (exoteric) and astronomical (esoteric) classification of months.

08. Distances of the planets from the Sun.

09. The cause of elliptical shape of orbits of the planets.

10. "Planets are fried on a skewer".

11. The reason of precession of the equinoxes.

12. The mechanism of cooling of the surface layers of a planet.

13. A gradual increase in the angle of inclination of the axis of rotation of planets.

14. The Earth's gravity decreases with time.

15. The speed of rotation of planets - what is the reason.

16. The causes of discrepancies continents and separation of Pangaea.

17. Dependence of intensity of solar radiation from the latitude of the sun.

18. The reason for existence of rings at the giant planets.

19. The final goal of the evolution of life on the Earth.

20. A commentary to the cosmological hypothesis.

21. The reason of absence of the atmosphere on the moon.

22. What is a comet?

23. Formation of planets in the solar system.

24. The reason of tides is not gravitation of the moon, and the pressure of the heated by the Sun atmosphere.

25. Cepheids are double stars.

DISPUTE BETWEEN THEORIES OF HEATING AND COOLING THE PLANET

A theme of global warming is one of the most popular issues currently being debated not only in scientific but also in the social circles. However, this issue is inextricably linked with the theme of ice ages. Yes, warmings and coolings because happening now in the world global warming is not the first in the history of the Earth. As well as the Ice Age was not the only one.

The theme of cooling and warming is closely intertwined with another scientific problem. A very long time (at least throughout the XX century) in science, there is a dispute between supporters of two concepts. Some people believe that the Earth cools, others – that it is heated. This question is really very difficult to analyze because both we have to consider at the same time an incredibly large number of astronomical, climatic, physical, chemical and biological factors. And besides in addition to the dispute between supporters of heating and cooling there is another division of opinion - within the concept of warming earth. Some scientists attribute the warming of the Earth to the action of only physicochemical factors - for example, of "heat generation due to gravitational compression of the planet". Others blame only "the human factor" – i.e. an increasing of "greenhouse effect" due to the growing in atmospheric levels of carbon dioxide in the combustion of various fuels.

Let's look who is right – i.e. is the Earth cool or heat, and what causes of one or the other. And we will say immediately - as it is often in science (and not only in science) the truth includes all (or almost all) existing points of view, at first glance, seeming contradictory. All the matter is that the Earth simultaneously heats and cools. Heating causes are as natural physicochemical factors, both natural and anthropogenic. Plus - biological.

After the Earth was formed from the solar material – it has been thrown out of the sun in outer space in the form of a giant drop of hot matter - it began to cool. The fact is that when the substance of the planet was in the depths of the Sun it was subjected to more significant transformation by gravitation much greater than it happens in the depths of the planets themselves.

So, the surface of arising planet first began to cool. Interior of the Earth (and of any other planet) cool down much less because to center the planet degree of transformation of elementary particles in chemical elements by means of gravitation increase.

So, the planets after leaving the Sun (star) start to cool, and primarily the surface layers exposed to the cooling. As you can see proponents of cooling the Earth are right - but only on half.

However, after leaving the Sun the planets begin to be exposed to such a powerful heating factor as elementary particles emitted by the Sun. Any star heats up "its" planets with the help of emitted elementary particles. The heating is provided in two ways:

1) Collisions of emitted particles with chemical elements in the composition of the planet;

2) Chemical elements of the planet accumulate (absorb) solar particles. More than half of reaching the planet solar elementary particles has Repulsion Fields. Most of them are the photons with Repulsion Fields belonging to the radio, micro frequency and visible ranges. Just this kind of particles with Repulsion Fields accumulated by chemical elements in the composition of the planet increase the total temperature of chemical elements.

These two ways of heating the particles of the planet by the solar particles differ from each other by the time during which the elevated temperature persists. We can consider a collision as a short-term way to raise the temperature. While accumulation (absorption) of particles with Repulsion Fields - it's not just a long-term way, and the way leading to the total heating substance of the planet not disappearing with time. It is this second factor has a major warming effect.

Solar particles meeting on the way the planet behave differently. They all follow influenced by the Field of Attraction of the planet. Some part of them is reflected immediately as a result of collisions with chemical elements of the planet. Other - remains in the upper layers of the atmosphere as a part of the ionosphere. Third - is absorbed by elements of all layers of the atmosphere. Fourth – is absorbed by elements of solid and liquid substances on the surface of the planet. However, ultimately the fate of most particles absorbed by the elements - to move from element to element, down toward the planet's center, subject to the action of its attraction.

By the way, the cooling of the surface layers of the planet, which happens every night, in cold weather, in cold climate and during the cold season caused by the fact that accumulated by elements solar particles with Repulsion Fields go down towards the center of the planet. And new solar particles at this time or do not come at all (at night), or come a little (in the cold season, in cold climates, and in cold weather). But this cold - it's just the return of chemical elements in the surface layers of the planet to their natural state, which is inherent for them out of the process of transformation by the solar particles.

It turns out that supporters of both concepts are right in their own way. Leaving the Sun the planets cooled though not completely. To the center - the temperature is maintained at a high level, although lower than it was when the substance was part of the Sun. However, the cooling of the planets is prevented the heating them by the particles with Fields of Repulsion emitted by the Sun. Solar particles "settle" in the first turn, in the center of the planets. The farther from the center, the less solar particles are accumulated in the elements, so to the periphery of the planets the temperature of substance decreases. Although we should not forget that to the center of any celestial body temperature rises due to the transformation by gravity.

People need to stop blaming others and themselves that someone or something disturbs them to exist. We just survive and adapt. We do not know for what we are doing this, but that is the whole point of our lives. The life is so mysterious. Each of us has own view of the world, but in general this knowledge is negligible. And we do not realize of what we are the part. We can say a million times - "God", "Creator". But this is not one iota reveal to us the essence of this Aught. So why we do not just go through life knowing that everything is subject to change. And if you're unhappy now, it does not mean that it always will be.

As you will soon know, the people - these are especial animals, and we are ordained our own path of development.

NUCLEI OF GALAXIES AND STARS - A COMPARATIVE CHARACTERISTIC

To begin with, in the center of any galaxy there is a celestial body. We call it as a Nucleus of the galaxy. The size of any nucleus of the galaxy is much larger than any star. Nuclei of galaxies are formed from material ejected from the bowels of even larger by size celestial bodies – Nuclei of Super galaxies.

The Nucleus of any Super galaxy first begets Nuclei of galaxies of the largest size – i.e. containing more chemical elements than others.

The greater is the number of chemical elements in the composition of a celestial body, the higher is the degree of transformation of the particles of elements of this celestial body - i.e. the greater is the overall temperature of the substance. Thus larger celestial bodies have the higher temperature compared with smaller. Of course it's provided that the original chemical composition of celestial bodies was identical. For example, the temperature of the Nucleus of any galaxy is always greater than the temperature of substance of any star generated by this Nucleus. Or the temperature of substance of any star is always higher than the temperature of even the largest of the planets. And the reason for that – is a smaller number of chemical elements in stars compared to galactic nuclei, as well as in planets compared to stars.

But let's back to where we started. The largest nuclei of galaxies having in their composition most of all of chemical elements generated first. And the explanation is following.

Any Galactic Nucleus in the most beginning of its life has in its composition more chemical elements than it has now. The more substance is in the celestial body, the higher is the temperature of the substance - i.e. in the most degree the particles of its elements are transformed – i.e. the faster the particles with Fields of repulsion emit Ether. And besides that in the initial stage of the life of the galactic nucleus its chemical elements were richer by particles with Fields of Repulsion. All this taken together leads to the fact that in the early Nucleus of the Galaxy chemical elements had the higher temperature – i.e. their mass was smaller and the antimass was more than later (and for example, now). And that's why more number of chemical elements was expiring from the Nucleus moving away from its center. Of this expiring substance the stars just were forming. And accordingly those earlier stars absorb more substance.

I.e. the stars that were formed first contained more chemical elements. And besides that the substance of earlier stars was more ornate by particles with Fields of Repulsion.

The same can be said of the nuclei of galaxies. Those of them that have arisen before contained a larger number of chemical elements. And chemical elements themselves were richer by particles with Fields of Repulsion. Therefore, the early nucleus of the galaxies were larger than latest – i.e. had in their composition more chemical elements.

To confirm these considerations we give the following facts.

There are two main types of galaxies: ***elliptical*** and ***spiral***. Here are their characteristics. "About 25% of studied galaxies have a circular or elliptical shape. Therefore they are called elliptical galaxies (in the classification this type of galaxies is denoted by E). These are the simplest systems in structure, star composition and nature of internal motions. In them there are not found of stars with high luminosity (supergiants), the brightest stars in elliptical galaxies - red giants. ... Depending on the degree of visible compression, elliptical nebulas are divided into eight subtypes: from the spherical systems E0 to lentiform E7 (the number indicates the degree of compression)" ("***Physics of Space***", the article "Galaxies", main editor - prof. ***SB Pikel'ner***).

"The other most common type of galaxies (about 50%) has very diverse structures. These star systems have two or more ragged spiral arms forming a flat area of the "disk", and in the center of the galaxy there is located a spheroidal core. They are called spiral and are denoted by S symbol" (ibid. – "Physics of Space", the article "Galaxies").

As is known, in spiral galaxies the blue giants are and they are located on the outskirts of these galaxies in their arms. Naturally, in these galaxies there are many of the red giants, which are closer to the nuclei of galaxies.

The heated substance that glows with blue light has a higher temperature than the incandescent substance that glows with red. It can be concluded that the blue giants are hotter than red. And *the temperature of big blue stars is explained by the large number of chemical elements in their composition, which automatically entails a larger value of the degree of transformation of particles in elements. Accordingly, the less temperature of red stars is explained by the lesser number of chemical elements in the composition of these stars and the lesser degree of transformation of particles.*

And now immediately let's turn to the analysis of galactic nuclei.

As was said above we can identify two main types of galaxies - elliptical and spiral. Elliptical have the shape of a ball or an ellipse, and spiral are lens-shaped with arms. The wide part of ellipse of elliptical galaxies is an area from which to further the arms will grow like in spiral galaxies (although not as big). And the arms of spiral galaxies and the thickening of the ellipse in elliptical galaxies are located in the same plane as the equatorial plane of the galaxy nucleus.

As is known, only red giants are observed among the stellar population of elliptical galaxies, and there no blue giants. What can tell us this fact? About that the Nuclei of those Galaxies that now have an elliptical shape initially contained a relatively small number of chemical elements (compared with the nuclei of spiral galaxies). It is a small number of chemical elements in their composition has not allowed them to have such large temperature of the material to throw out from themselves a large number of chemical elements. And as a result stars formed from material ejected by the Nuclei of such of galaxies did not contain initially so many chemical elements in order the temperature of the substance of these stars matches the blue luminosity. While in the spiral galaxies blue giants are much enough and they are located as has been said on the outskirts of these galaxies in their arms. This means that those nuclei whose Galaxy now have a spiral shape, originally had in their composition enough chemical elements to produce stars with a large content of substance. And this led eventually to a greater degree of transformation and to the blue luminosity. ***That's why blue giants are in the spiral galaxies.***

Now let's talk about the relationship between the shape of galaxies, the numerical composition of their nuclei and the age of them.

So, we found that ***spiral galaxies are older and elliptical - younger.*** This means that any Galaxy early in his life had a circular shape. Then its shape gradually begins more and more to resemble a lentils. And in the future the Galaxy gradually transforms into the flat disc with arms. I.e. elliptical galaxy becomes a spiral. Obviously, the lentiform shape indicates the beginning of the formation of a flat disc with arms. So why have more ancient galaxies which we call the spiral a flat disc with arms and younger - elliptic - this or not at all (round) or a flat disc is in the bud (lentiform)?

Here is an answer to this question. Fact is that at the center of any galaxy there is a celestial body – a Nucleus of Galaxy. All Nuclei of Galaxies are generated by one or another Nucleus of Supergalaxy. And a Nucleus of Supergalaxy as any celestial body larger than the planet emits elementary particles. These elementary particles, reaching Nuclei of Galaxies, are accumulated in the substance of Nuclei (on the surface of chemical elements and in empty space between them). Between elementary particles emitted by Nuclei of Super galaxies (and by any other celestial bodies), the particles with Fields of Repulsion (red) predominate. Thus, there is an increase of the total Field of Repulsion of Nuclei of Galaxies.

Nuclei of Galaxies rotate around their axes as well as planets and stars. And the reason for this rotation is the heating from the side of the generated them celestial body. In this case, the Nuclei of Galaxies are heated by radiation of the Nuclei of Supergalaxies generated them. At any Nucleus of Galaxy (as well as at stars and planets) their rotation axis is perpendicular to the straight passing through the center of the Galaxy Nucleus and the center of the Supergalaxy Nucleus generated it. It is for this reason the area of equator (of a Nucleus of Galaxy or of a star or of a planet) is heated by radiation in the greatest extent – i.e. accumulates the most amount of free particles with Repulsion Fields. In this case, the poles of Nuclei of Galaxies (or stars, or planets) are the least heated areas. As a result, the total gravity field of equator is the smallest, and of the poles is the largest. With the passage of time until the Nucleus of Galaxy is moving around the Nucleus of Supergalaxy and continues to accumulate free particles, this imbalance *"equator/pole"* has been increasing. The Field of Attraction at the equator more decreases and temperature of the substance, on the contrary, increases. An increase in temperature of the substance at the equator leads to the fact that it is this region of the Galaxy Nucleus throws out of itself most of all substance. And of this substance in the future the stars will arise that form the silhouette of the Galaxy.

Why do the stars formed from ejected material, tend to align along the same plane (form a flat disc) corresponding to the equatorial plane of the Galaxy Nucleus? Yes, because at the same time the total Field of Attraction of both poles effect on these stars. More specifically, not only of poles but also of entire substance of each of two hemispheres on each side of the equatorial plane. The Nucleus of Galaxy, as any celestial body formed from the incandescent material has a very symmetrical shape. One hemisphere is almost 100 % identical to another. As a result, the values of total Attraction Fields of both hemispheres are equal. And the stars, ejected from the Galactic Nucleus, "feel" this. I.e. **they are equally controlled by the attraction of each of two hemispheres. That's why the stars tend to align in the form of a thin disk along the equatorial plane of a Galaxy Nucleus.** If it was not – i.e. there would be no "control" as an action of Attraction Fields of both hemispheres - all Galaxies would have only a circular shape, there would be nor lentiform, nor spiral.

THE MECHANISM OF ROTATION OF PLANETS

Before we will talk about the reasons that cause the planets to rotate around their own axis, let us recall some features of their structure.

Dense and liquid part of any celestial body of the planetary type manifests outside an Attractive Field. This is explained by the fact that among the chemical elements of dense or liquid substances there are predominantly those that exhibit outwards Fields of Attraction. And the value of these Fields of Attraction is the highest in comparison, for example, with the same liquid substance.

As we already know, the value of any Field (and of Attraction and of Repulsion) decreases with distance. This means that during the distancing from the solid or liquid surface of the planet, the value of its Fields of Attraction decreases. This means that there occurs the decrease of the Force of Attraction to the planet occurring in other bodies, and makes them strive to fall on the planet. That is why, the farther from the surface of the planet a spaceship is moving away, the smaller is the value of the Force of Attraction amount occurring in it to the planet – i.e. the less it "feels" its attraction.

As for the chemical elements of the atmosphere of the planet, they exhibit outward or very small Fields of Attraction, either neutral or Fields of Repulsion. If the elements-gases accumulate on their surface free solar particles (among which there is a domination of the particles with the Fields of Repulsion), then they do not exhibit outward any Field of Attraction. Thus, **the elements of the atmosphere, being even slightly heated by the accumulation of particles with Fields Repulsion begin to demonstrate outward Fields of Repulsion – i.e. they emit Ether (energy)**. And as a result, the chemical elements of the atmosphere shield the chemical elements of the dense and liquid part of the planet, reducing the outward manifestation of the Gravity Field of the planet.

However, note your attention, none of the celestial body (including the planet) doesn't exhibit outward Field of Repulsion itself, without heating. Only Field of Attraction. And Field of Repulsion appears only as a result of the heating from a celestial body that generates it. In the case of the planets, they are heated by the stars generating them.

It is the appearance at a heavenly body a Field of Repulsion is the cause of its rotation about its own axis. In particular, that's why the planets and stars rotate.

Star emits elementary particles. The particles with Fields of Repulsion predominate among the emitted particles. When the star emits these particles, they move by inertia until they reach any of the planets. There, they are accumulated on the surfaces of the elements in the chemical composition of these planets. And this process of accumulation begins from the elements of the atmosphere. Namely atmospheric elements play a major role in a forming of the Field of Repulsion of the planet. Point is that, as already mentioned, the chemical elements of the atmosphere exhibit outward or weak Field of Attraction, either neutral, or even Field of Repulsion. So, when these elements accumulate on the surface solar particles (among which the particles with Fields of Repulsion predominate), they begin to exhibit outward the Field of Repulsion.

Moreover, the solar particles themselves are accumulated in the upper layers of the atmospheres of planets, in the ionosphere, held by the action of the total Field of Attraction of the planet.

As a result, in those regions of the atmosphere (or of the surface as on Mercury), which are currently turned to the star (i.e. lighted by it), the whole totality of accumulated particles with Fields of Repulsion forms the total Field of Repulsion. This Field of Repulsion is an "*<u>ethereal shield</u>*". Namely this "shield" hinders the process of rapprochement of the planet with the star. In other words, ***namely this arising Field of Repulsion (emitted ether) just also and prevents to drop of the planets on the star - for example, on the Sun! If the Sun did not emit elementary particles, all the planets would have fallen on it (on the Sun)***. As also in the area of the planet, turned at the moment to the star (i.e. it is located from the center of the star at a minimum distance) there arises a Field of Repulsion (ethereal shield), this area begins to move away from the star. And there is no distancing of the whole planet from the star. No, instead this, only the heated area turns away from the star. The neighboring area of the planet that all this time was on the night side and is cooled, so do not have the same ethereal shield. This means that this area exhibits outwards the Field of Attraction – i.e. absorbs ether. And so this area gravitates to the star and seeks to be closer.

It turns out that *at the same time as the heated area tends to move away from the star, the area cooled on the night side, by contrast, tends to fall on the star. And finally there occurs a rotation of the planet around its own axis. And this is constantly.* As soon as the Field of Repulsion (ethereal shield) arises in the heated region of the planet, the planet rotates, substituting to the star to heat another "side".

That's how we can briefly describe the mechanism of rotation of any planet in the composition of our solar system, or in any other solar system with another sun.

Stars also rotate around their axis, since they are heated by the radiation of the Nucleus of the Galaxy. Nuclei of Galaxies are also rotate due to the heating by Nuclei of Supergalaxies.

ON WHAT DOES THE HEATING OF PLANETS BY STARS, OF STARS BY GALACTIC NUCLEI, OF GALACTIC NUCLEI BY NUCLEI OF SUPERGALAXIES AFFECT?

The heating of the planets by the Sun - this is the only factor determining such astronomical characteristics of the planets as their rotation around their own axis, the slope of the equatorial plane to the ecliptic plane, the periodic change in the distance between the hemispheres of the planets and the center of the Sun, as well as the distance between the center of the planet and the center of the Sun.

The same can be said about the dependence of the same characteristics of the planets from their heating any other star (besides the Sun), around which they rotates.

Also, the similar astronomical characteristics of stars (rotation, tilt, the change of the distance of hemispheres and also the distance from center to center) are caused by the heating of them by particles emitted by the Nucleus of Galaxy that generated them. The same characteristics of Galactic Nuclei are caused by the heating of them by the Nuclei of Supergalaxies that created them.

All celestial bodies, formed from material ejected out of the bowels of other celestial bodies, are spherical. This is explained by the fact that the substance ejected from the bowels of the heavenly bodies, is in the molten state. The spherical shape of the body allows using the space most economically during the process of compound of elementary particles and chemical elements into single whole. A red-hot state of substance tells us that all (or almost all) of chemical elements in the composition of this substance have Fields of Repulsion. Namely this allows them to move freely relative to each other when they take up free places, submitting to action of the centripetal Field of Attraction of the formed celestial body. This information relates to large satellites, planets, stars, nuclei of galaxies and nuclei of Supergalaxies. The Central Celestial Body of the Universe is not formed by ejection from other celestial body. It was born during the process of unification of elementary particles in the first period of existence of this Universe.

The heating of celestial bodies by falling particles is realized, at first, at the expense of the process of collision of particles with chemical elements. And secondly, this is due to the accumulation by chemical elements of a celestial body of elementary particles with Fields of Repulsion. Moreover, namely the second factor - the accumulation of particles - allows absorbing elements retain their "received" temperature. At the same time the degree of transformation increasing during the process of collision quickly returns to the previous level.

Accumulation of elementary particles by a celestial body is carried out thanks Fields of Attraction in separately taken elements, but mainly due to the presence in any celestial body of the Field of Attraction. In this case, all particles falling on the bombarded celestial body initially absorbed by the chemical elements of the surface layers of the celestial body (except those that are rejected). But then, all particles absorbed by elements "flow" down toward the center of the celestial body.

Under the rays of the "heating" celestial body at any given moment of time is always the whole hemisphere of the "heated" celestial body. We are not talking about the northern or southern hemisphere. It's just a hemisphere bombarded by particles. So, all elements of surface layers of the "illuminated" hemisphere "accumulate" particles for the entire celestial body, and especially for its central part - for the kernel. Thus, as an illuminated celestial body would not change its position in the space, elements of its central part continue to accumulate particles with Fields of Repulsion, and to be heated by such way.

The substance of celestial body disposed in its equatorial plane is heated most of all. And this is due to the fact that namely the region of equator in the early life of a celestial body was closest to the center of the heating it (and gave rise to it) celestial body.

1) Arising of the Field of Repulsion in the region of the planet that faces in the given moment to the sun, due to accumulation by surface layers of elementary particles emitted by the sun is the cause of the ***planet's rotation*** (and of any "heated" celestial body);

*2) **The distance between the center of a celestial body and the center of its parent heavenly body*** is entirely due to the total temperature of the substance located in the equatorial plane of a considered celestial body;

*3) **The constant change in position of the axis of rotation of a celestial body about the real or imaginary axis of rotation of its parent celestial body*** (such as the planet's rotation axis relative to the axis of rotation of the Sun) is due to the periodicity of the heating and cooling of each of two hemispheres. On the imaginable axis of rotation we should speak in those cases when a celestial body does not rotate. For example, the Central Sun of the universe does not rotate;

4) The axis of rotation of a celestial body is constantly in the process of changing of *the angle of inclination to the ecliptic plane*, which is due to the emergence of a constantly acting Field of Repulsion.

REASONS FOR BEGINNING OF THE ROTATION OF PLANETS

The rotation of planets, which seems so natural, was not peculiar to the planets immediately after they arose. Special conditions were required for it started.

Planets are formed from material ejected from stars. The temperature of stars is low compared to the temperature of Nuclei of Galaxies and Supergalaxies, especially compared to the Central Sun of the universe, and this is because of the smaller number of chemical elements in their composition (which reduces the degree of transformation caused by gravity). Therefore, a significant amount of material sufficient for the formation of planets is ejected by stars only in the plane of the equator. Namely in the plane of the equator of a star there is an accumulation of the most number of particles with Fields of Repulsion emitted by the Nucleus of Galaxy, gave birth to this star. At the moment of birth - ejection from the sun - the planet can be with whatever side of the Sun in relation to the Nucleus of Galaxy. A planet necessarily begins to rotate some time after its birth. But for the beginning of rotation it is necessary that the planet does not shield by the sun from the action of gravity of Galaxy nucleus or is not on the line connecting the Sun with the Galactic Nucleus. In any other position the planet necessarily starts to rotate. However, planets after their occurrence begin to revolve around the sun – i.e. move around in a circle. Therefore, even if at the moment of time the planet was located on a straight line drawn through the centers of the Sun and the Nucleus of galaxy (behind or in front of the Sun), due to the orbital motion the planet after a while will go to the position "laterally".

All stars emit elementary particles. Therefore after formation the planets begin to be bombarded and heated by the sun. But only that hemisphere is heated that faces the sun. At the same time, the other hemisphere, the opposite of what turned to the sun is not heated and therefore turns colder. And accordingly, the total Field of Attraction of the night hemisphere, not facing the sun, is larger compared to the heated hemisphere.

So, at the planets at the moment of their location "laterally" of the Sun that created them their not heated hemisphere, located on the night side, is attracted by the Nucleus of Galaxy. The Force of Attraction, caused by the Nucleus of Galaxy, due to the large distance to it is less than Forces of Attraction caused by the Sun. But, anyway, this Force of Attraction exists and exerts its influence on all celestial bodies in the solar system. Simultaneously with this the heated hemisphere of the planet begins to move away from the Sun. And here it is the gravitation from the side of the Galactic

Nucleus decides "the fate of the planet". Or rather, this attraction is a reason for starting rotation of the planet. I.e. as a result, the planet revolves around its own axis, as the hemisphere opposite heated, colder, tends to move in the direction of the Galaxy Nucleus, and the heated hemisphere moves away from the Sun.

The phrase "laterally" we hope you guessed it, means that the planet is not shielded by the sun, gave birth to it, from the Nucleus of Galaxy, and is not on the line connecting the centers of the Nucleus of Galaxy and the Sun.

RETROGRADE AND PROGRADE ROTATION OF PLANETS

Due to astronomical observations, we know that most of planets in our solar system revolve in the ***prograde direction*** – i.e. counterclockwise. And this direction of rotation coincides with the direction of rotation of the Sun. However, two planets in the solar system rotate in the ***retrograde direction*** – i.e. clockwise. Venus and Uranus rotate by such way.

Let's examine why not all planets of the solar system rotate in the same direction.

As already mentioned, the reason for starting rotation of each planet is an action of two factors – a tendency of the hemisphere heated by the star (the sun), to move away from it and an attraction of the opposite, cooler hemisphere by the Nucleus of galaxy. As already mentioned, the rotation of the planet began only when the planet was located "laterally" of the Sun (star) to the Nucleus of the Galaxy. So, prograde or retrograde the rotation of the planet became, it depended on only one factor. And namely from which "side" of the Sun the planet was located at the start of the rotation. We can conditionally defined designate one "side" of the Sun as a right, and other - as a left. For example, if you look at the Nucleus of Galaxy from the position of an observer on the Sun, then the "side" of the Sun on the right is the right and one on the left is the left.

So, if the planet at the start of rotation was on the right "side" of the Sun, then it began to rotate counter-clockwise – i.e. forward. Most of the planets in our solar system were in such situation. If the planet was located on the left "side" of the Sun, then it began to rotate in a clockwise direction – i.e. in the opposite direction. Venus and Uranus were in this strategy.

But why the planets did not change their direction of rotation after they were in the course of rotation around the sun on its other "side".

And here's why.

The magnitude of the Force of Attraction that arises in any planet or satellite in the solar system composition in relation to the Nucleus of Galaxy is always less than the Force of Attraction that arises with respect to the Sun (i.e. to a star). And the reason for this is the difference in distances. The Nucleus of the Galaxy is very far. And so, even despite of its huge size (much larger than of the Sun), the magnitude of the Force of Attraction arising in relation to it, is less.

When the planet did not yet rotate, one its hemisphere completely turned to the sun, and another was completely away from it. This means that the turned away hemisphere did not "feel" attraction of the Sun (precisely because it turned away from it). There was only attraction of the Nucleus of the Galaxy. But once the heated hemisphere began to turn away from the Sun, starting thus the planet's rotation, at the same time the colder, turned away hemisphere began gradually to come back to the illuminated side. And once that happens, the Force of Attraction directed toward the Sun begins to act on it, and it is greater than the Force of Attraction to the Nucleus. As a result, after the rotation of the planet began, its direction does not change. And all because of that now, all time, when chilled on the night side area begins to come back to the illuminated side, the Gravity Field of the area forces this area to move in the direction of the Sun. And so, there is a rotation of the planet. We recall that on the illuminated side of a planet a Field of Repulsion is forming, and that, in fact, makes the heated area to move away from the Sun.

As you understand, we can talk about the prograde and retrograde rotation not only of planets, but also of stars and Galactic Nuclei.

THE SEASONAL (EXOTERIC) AND ASTRONOMICAL (ESOTERIC) CLASSIFICATION OF MONTHS

The modern world lives according to the Julian calendar, according to which in the year there are 12 months, of which three belong to the winter season, 3 - to the summer, 3 – to the autumn, and 3 - to the spring. And, for the northern hemisphere months relating to any of 4 seasons are specularly opposite to the months characterizing the same seasons in the southern hemisphere.

But it is well-known facts. And now let us turn to what science does not know.

Usually people classify months in accordance with the annually recurring processes of a general lowering of the temperature of the surface layers of the Earth (autumn, winter), and of a general increase in temperature (spring, summer). In autumn and winter the earth's surface is gradually cooled and in spring and summer - is gradually heated. September, October and November attributed to the autumn months, to winter - December, January and February. March, April, May belong to spring. To summer - June, July and August. This is an exoteric systematization. It is known even to younger students.

With regard to the esoteric classification of the esoteric, it is known only to occultists, and even then not for everybody. Anyway, people studying in the Trans-Himalayan Esoteric School are aware of its existence and use this information in the course of their meditations.

We will tell you more details.

If we classify months in accordance with astronomical observations, then we should divide them into four groups of three months in each. Three months would be attributed to the winter solstice, three - for the summer, three - to the autumnal equinox, and the last three - for the spring.

Moreover, the days of solstices and equinoxes themselves would represent central "cutoffs" in each of the four groups.

But at the same time there would be the next problem. The days of solstices and equinoxes are not located in the middle of the month - all they are in the early 20's numbers. I.e. we should shift the beginning of all months so that the days of solstices and equinoxes would come about on the 15th numbers of the months.

But humanity keeps a calendar relying not on astronomical observations and in accordance with the processes of annual change of heat and cold on the earth's surface.

Why do not coincide with each other the astronomical (esoteric) and seasonal (exoteric) classification of months?

For the northern hemisphere the winter season can be correlated with the time of the winter solstice, the summer season - with time of the summer solstice, autumn - with the period of the autumnal equinox and the spring - with the period of the vernal equinox. For the southern hemisphere a correlation of the seasons and moments of solstices and equinoxes will be specularly opposite.

We remind you that the ***solstice*** (summer or winter, it does not matter) - is the time when one of the hemispheres of the most turned to the Sun (close to it) and other at this time maximum turn away (far from it). At the same time, the ***equinox*** (spring or fall) is the time when both hemispheres are from the Sun at the same distance

However, *for each of the hemispheres, each of the seasons "lags behind" more than a month against a group of 3 months, belonging to one or another solstice or equinox*. For example, February for the northern hemisphere - this is the last of the winter months. But from the astronomical point of view, February is the first month referring to the group of vernal equinox months. Or, for example, May - for the northern hemisphere is the last month of spring. But from the astronomical point of view this is the first month referring to the group of summer solstice months. Similar information we can lead to August, and November. August in the northern hemisphere - this is the last month of summer. But with the astronomical positions - this is the first month belonging to the group of the autumnal equinox. November is the last month of autumn. And in astronomical terms - this is the first month of the winter solstice.

What is the cause of this "displacement" of the seasonal classification regarding astronomical? The whole point is in the features of the heating and cooling of the surface and intermediate layers of the planet by solar particles.

In the article "Planets are fried on a spit" we will discuss in detail the reasons forcing the hemispheres of the planets periodically change the distance to the center of the Sun. We will talk about the main points.

As part of any planet we conditionally have allocated a core, surface layers and layers, intermediate between the core and surface layers. However, it is really just a convention. No boundaries between the layers, one layer smoothly passes into another. In the composition of surface layers and in the composition of intermediate and in the core there are layers, located near to the center of the planet, and there are layers located farther from it. Solar photons falling on the planet at first are accumulated (absorbed) by chemical elements of the overlying surface layers. And then, under influence of the planet Field of Attraction "gravitate" into the layers located below. And so they gradually move from one layer to another, farther to the center. As a result, in the planet's core the concentration of solar photons is the largest. Only particles with Fields of Repulsion are able to raise the temperature of the chemical elements absorbing them. The photons with the Fields of Repulsion predominate in the composition of the solar radiation reaching any of the planets. That is why an accumulation by planets of solar particles leads to the total increase in temperature within the planet. The process of "settling" of photons from more superficial layers into the underlying takes certain time – i.e. it does not occur instantaneously.

And now we will talk directly on why the seasonal classification of months is displaced relative to the astronomical classification.

In general it can be said that the weather of each month depends firstly on the total amount of solar photons received at a given time by the earth surface, and secondly, it depends on the degree of heating of the lower, deeper layers of the planet.

The reason for changing the distance between the hemispheres and the center of the Sun is a change of the total temperature of the deep surface and also of intermediate layers in each of the hemispheres.

The moment of the winter solstice in the northern hemisphere is time of the highest total temperature of deep superficial and intermediate layers. This time is the turning point. Since then, this temperature begins to fall more, which leads to increase in the total Field of Attraction of the hemisphere as a whole. Because of this, gradually decreases its distance from the Sun.

From the moment of the summer solstice to the winter temperature of deep superficial and intermediate layers of the northern hemisphere is growing due to the accumulation of solar photons (solar energy). And as the temperature increases, the hemisphere gradually moves away from the Sun. Seclusion of the hemisphere leads to the fact that the superficial layers receive less and less solar photons. The Earth's surface cools more and more - do not forget that the underlying layers due to their gravitation constantly take away photons.

Thus, during the time of the summer solstice to the winter in the intermediate and superficial layers of the northern hemisphere opposite processes occur. The Earth's crust is gradually cooled, and intermediate layers are heated more and more.

Underlying superficial and intermediate layers are heated for the simple reason that they receive solar photons from the upper surface layers. It takes time to the "settling" of photons from the surface to the intermediate. Later the photons of the intermediate layers gravitate to the center of the planet. ***Settling of photons leads to the cooling of layers where from they are moving downwards, and to the heating of layers in which they are coming down.***

It turns out that in the period from the summer solstice to the winter, due to an approximation of the northern hemisphere to the sun, chemical elements of the earth's crust in this hemisphere receive more and more solar photons. In the future, these photons will start their way down, coming down into the intermediate layers in the period from the summer solstice to the winter.

This is the essence of explanation why the greatest heating of the intermediate layers of the northern hemisphere occurs in the period from the summer solstice to the winter. Please note - on the surface of the planet at this time is getting colder, and in depth - warmer and hotter.

Do not forget that even when the hemisphere is increasingly turning away from the Sun, its crust continues to accumulate solar photons, although in much smaller quantities.

Photons need time to move through the substance of the planet down to its center. Moreover, the scale of the surface layers of the earth within which we usually estimate the temperature of Earth's surface substance and talking about warming or cooling is very small in comparison with the scale of the entire planet. We can assume that the humanity "does not plunge" into the planet generally deeper than 1 kilometer.

In the period from the summer solstice to the winter, when the northern hemisphere is increasingly turning away from the Sun, its surface layers receive less and less solar photons.

They are deposited down with a delay. I.e. to ensure that the photons passed from the surface layers to the intermediate, takes time. The process of "settling" of photons that fall into the surface layers during the period from summer to winter will continue after the winter and will pass to the spring. And as from summer to winter the amount of solar radiation received by the planet gradually decreased, and an amount of accumulated photons in the crust is little, the intermediate layers in the process of settling also get them a little, and the temperature decreases. As a consequence - the northern hemisphere cools. And as it cools, then its Field of Attraction increases. And it approaches to the Sun. And the most chilling accounts at the time of the summer solstice – in this time into the intermediate layers there occurs the settling of photons accumulated by the crust at the winter solstice, and there were very little of them.

It turns out that the planet is delayed by six months in its response to the heating of the surface by solar photons.

And now we again will talk about why the seasonal classification offset relative to the astronomical classification of months.

The moment of the summer solstice corresponds to the minimum distance from the North Pole to the center of the Sun. At this time, the surface layers of the Earth's surface of the northern hemisphere receive the greatest amount of solar particles with Fields of Repulsion (as well as of particles with Fields of Attraction). After the moment of summer solstice the earth's surface begins to be heated less and less.

Then why, oddly enough, *July* is the hottest month, and *August* is also hot enough? The thing is that by this time the deep layers of the surface layers of the planet are already warmed enough. I.e. these layers have received enough photons accumulated by the most superficial layers during the period from the winter solstice to summer. Warming of the deep layers in the composition of the surface layers of the earth is the cause of reducing the Field of Attraction of the planet. For this reason, during the night time atmosphere, hydrosphere and solid surface of the crust do not so much cool, as it happened if deep layers would not be heated. We recall that the cooling of substances on the earth's surface is due to "settling" down of accumulated photons with Fields of Repulsion – i.e. due to their moving toward the center of the planet.

Thus, because of the not so strong cooling of the earth's surface at night, *July* is the hottest month (despite the fact that it is the last month, referring to the period of the summer solstice). And *August* by the same reason does not be attributed to the autumn, but to the summer season.

At the same time, *May* refers to the spring season, and not to the summer, and just because of the fact that by this time the deeper layers in the composition of the surface layers still not heated enough, so at night the earth's surface cools to a greater extent than that observed, for example, in *August*. While at the same time, May refers to the group of months of the summer solstice and the earth's surface at this time already receives enough of solar particles.

Similar information we can lead to other three seasons.

Аналогичную информацию можно привести для остальных трех сезонов.

The fall season as it is known consists of three months - September, October and November. But only two of them - September and October are really the autumn. In *September*, the vernal equinox occurs. And *October* immediately follows September. The first month of those, relating to autumnal equinox is *August*. However, as mentioned, due to the high temperature of the surface layers of the Earth, it belongs to the summer season.

As for *November*, then it is not yet so cold to attribute this month for the winter season. Although namely November is the first month of those, that "surround" the winter solstice. November is the first true winter month. But this is only by the esoteric classification.

In exoteric systematization December opens winter. But as you know, the winter solstice is in December. On December 22 in the northern hemisphere there is the longest night and the shortest day. So according to the esoteric classification, **December** is the second month of winter, but not the first.

January is the second month of winter. But in reality it is the last. And February traditionally closing the winter season, in fact, opens the spring. It belongs to the vernal equinox. It is the first of three spring months of the esoteric classification. However, due to the fact that at this time the northern hemisphere is still quite far away from the Sun, and its surface is still cold (receives a little of solar photons), February not for nothing is attributed to winter.

In late March, the vernal equinox occurs. **March** is the second month of spring, according to the esoteric classification, but the first in accordance with the Julian calendar.

April is the last month, referring to the period of the vernal equinox. But according to tradition, this is the second month of spring.

And here we are again back to **May**. It is the last month of spring, as is commonly believed. But it is the first month of the summer solstice in accordance with astronomical observations. In May the surface layers have warmed up not enough that people view it as the summer season.

DISTANCES OF THE PLANETS FROM THE SUN

Mercury - 58 million km;

Venus - 108 million km;

Earth - 150 million km;

Mars - 228 million km;

Jupiter - 778 million km;

Saturn - 1.43 billion km;

Uranium - 2.87 billion km;

Neptune - 4.5 billion miles;

Pluto - 5.95 billion miles.

The distance of planets to the center of the star (spawned them), as well as their rotation, is related to the formation of the Field of Repulsion (ethereal shield) in the heated area of the planet. However, unlike the speed of rotation of the planet, the distance is caused not by to heating rate and directly by the magnitude of the Field of Repulsion arising in response to the heating by solar radiation.

The larger this magnitude, i.e. more is the rate of emission of ether, the farther from the star a planet will be located.

All planets in our solar system have now one or another angle of inclination of their rotation axis to the ecliptic plane. Inclination of the axis of rotation is due to the fact that the substance of the planet warms by solar radiation (accumulates particles with Fields of Repulsion). Substance in the equatorial plane of the planet warmed up to the greatest extent. This is explained by the fact that at the beginning of life on the planet, when it just started to rotate, the axis of rotation is perpendicular to the line drawn through the center of the Sun and the center of the planet. So, the inclination of the axis indicates that at the planet in the equatorial region and adjacent areas there is formed a continuously existing Field of Repulsion. The Field of Repulsion of a planet is formed by particles with the same Fields accumulated on the surface of chemical elements of the atmosphere - lingering in the composition of the surface layers of the planet. Actually, it is because of this constant Field of Repulsion planet "lean".

For what it's been said? Yes to the fact that when we talk about a Field of Repulsion (ethereal shield), the value of which determines the distance of the planet to the Sun, then we are talking about a Field of Repulsion that exists in the planet's equator, as it is the largest in magnitude.

I.e. *the magnitude of exactly this Field of Repulsion arising in the equatorial plane, and just will allow to estimate the distance between the planet and the sun. The greater the magnitude of this field Repulsion – i.e. the higher is the rate of emission of ether by the equator of the plane, the greater the distance between the planet and the Sun.*

The planets in the solar system can be likened to balloons. The air inside the balloon dome is heated by the burner flame and thus the balloon is moving away from the planet's surface (i.e. from its center). Only in the case of planets the Sun itself plays the role of the burner.

Let's think about why all planets move away from the Sun.

First, we should remind ourselves that the farther away from the Sun, the smaller the number of solar particles reaching this point.

So, during the heating of the substance of equator region, the Field of Repulsion of this area is growing. Due to the fact that solar particles accumulated in the surface layers linger there longer. Increase of Field of Repulsion says that the rate of emission of ether becomes greater than the rate at which ether moves to the Sun (i.e. greater than the sun's Gravity field at a given point). If the planet emits ether faster than it is attracted by the sun, it begins to move away. However, as already mentioned, with increasing of the distance from the sun there is a decrease of solar particles reaching the planet. This means that the rate of distancing of any planet from the Sun gradually decreases with increase of its distance from the Sun. I.e. the distancing of planets slows down, as they move away. We can say that there is a feedback mechanism. The farther from the Sun, the less of solar particles reach the planet, and due to this the magnitude of the Field of Repulsion formed by the planet decreases. As a result, there is no fast departure of the planet from the Sun. No, all planets move away gradually, slowly.

THE CAUSE OF ELLIPTICAL SHAPE OF ORBITS OF THE PLANETS

The reason for which the planets move in orbits having an ellipse shape is very simple. Planets are rejected by attraction of the Galactic Nucleus. The Nucleus of the Galaxy is a celestial body, much larger than any star. Namely it gave rise to all stars of our Galaxy, ejecting the substance from itself.

Aphelion of orbit indicates the direction exactly on the Galactic Nucleus.

"PLANETS ARE FRIED ON A SKEWER"

Due to astronomical observations, we know that every planet in the solar system rotates around its own axis. And we know that all planets have one or another angle of inclination of the rotation axis to the ecliptic plane. It is also known that during the year each of two hemispheres of any planet changes its distance from the Sun. But by year-end a position of the planets relative to the sun is the same as a year ago (or rather to say, almost the same). There are also such facts that astronomers do not known, but nevertheless, that exist. For example, there is a constant, but smooth changing in the angle of inclination of the axis of each planet. The angle increases. And, in addition, there is a constant and gradual increase in the distance between the planets and the Sun. Is there a connection between all these phenomena?

Answer - yes, of course. All these phenomena are due to existence in the planets as Fields of Attraction and Fields of Repulsion, due to the characteristics of their location in the composition of the planets, as well as changes in their value.

We are so accustomed to the knowledge that our Earth rotates on its axis, as well as to the fact that the northern and southern hemisphere of the planet for a year then move away, then approach the Sun. And with the rest of the planets everything is the same. But why planets behave this way? What motivates them?

Let's start with the fact that any of the planets can be compared with an apple impaled on a skewer and roasted over a fire. The role of "fire" in this case performs the sun, and a "skewer" is an axis of rotation of a planet. Of course, people often roast meat, but here we look at the experience of vegetarians, because fruits often have a rounded shape, which makes them similar to the planets. If we fry an apple over a fire, we do not move it around the source of fire. Instead, we rotate the apple, as well as change the position of the skewer relative to the fire. The same thing is happening with the planets. They rotate and change the position of the "skewer" throughout the year relative to the sun, warming, so their "sides".

The reason that the planets revolve around their axes, as well as throughout the year the poles periodically change the distance from the Sun, is about the same according to which we rotate the apple over the fire. The analogy with a skewer was not chosen randomly. We always keep the least deep fried (the least-heated) area of an apple over the fire. Planets also always tend to turn to the sun by their least heated side, the total gravity field of which is the greatest compared to other areas. However, the expression "tend to turn" does not mean that the way it really happens. The trouble is that any of planets at the same time has just two sides, the gravity of which to the Sun is greatest. This is the poles of the planet. This means that from the moment of birth of the planet, both poles simultaneously sought to take a position to be closest to the Sun.

Yes, when we talk about the attraction of the planet to the Sun, it should be noted that different areas of the planet are attracted to it in different ways, i.e. in varying degrees. At least - the equator. To the greatest extent – the poles. Note - the number of poles is two.

I.e. if two areas tend to be on the same distance from the center of the Sun. The poles throughout the existence of the planet continue to balance, constantly competing with each other for the right to occupy a position closer to the Sun. But even if one pole temporarily wins and gets closer to the Sun than other, this other, continues to "graze" it, seeking to turn the planet so as to be closer to the luminary. This struggle between two poles directly affects the behavior of the whole planet. It's difficult for the poles to approach to the Sun. However, there is a factor facilitating their task. This factor is the existence of the angle of inclination of the rotation axis to the ecliptic plane.

However, in the beginning of life planets had no inclination of the axis. The cause of emergence of inclination is the attraction of one of the poles of the planet by one of the poles of the Sun.

Let's consider how inclination of the axes of rotation of a planet appears.

When the substance from which the planets formed is ejected from the Sun, an ejection does not necessarily occur in the equatorial plane of the Sun. Even a small deviation from the equatorial plane of the sun results in that the formed planet is located to one of the poles of the sun closer than to another. And to be more precise, it is only one of the poles of the formed planet is closer to one of the poles of the Sun. For this reason, namely this pole of the planet experiences greater attraction from the pole of the Sun, to whom it was closer.

As a result, one of the hemispheres immediately turned toward the Sun. **So an initial inclination of axis of rotation appeared at the planet.** That hemisphere, which was closer to the Sun, respectively, immediately began to receive more solar radiation. And because of this the hemisphere from the outset was heated to a greater extent. Greater heating of one hemisphere of the planet becomes the reason that the total gravity field of this hemisphere decreases. I.e. during the heating of the hemisphere approached the Sun began to decrease its tendency to move closer to the pole of the Sun, attraction of which made the planet to be closer. And the more this hemisphere warms, the more leveled the gravitation of both poles of the planet, of each to "its closest" pole of the Sun. As a result, the heated hemisphere turns away from the sun more and more, and the more chilled begins to approach. But note, as this change of poles happened (happens). Very peculiar.

After the planet was formed from material ejected by the Sun, and now revolves around, it immediately begins to be heated by solar radiation. This heating makes it rotate around its own axis. Initially an axial tilt was not. Because of this, the equatorial plane is warming to the greatest extent. Because of this, namely in the equatorial region not disappeared Field of Repulsion appears in the first turn and its magnitude is greatest from the outset. In areas adjacent to the equator with time not relieved Field of Repulsion also appears. A value of the squares of the areas, on which there is a Field of Repulsion, is demonstrated by an angle of inclination of axis.

But the sun also has a constantly existing Field of Repulsion. And, like the planets, in the area of the Sun's equator the value of its Field of Repulsion is greatest. And since all planets at the moment of ejection and formation were located in the Sun's equator, they revolved thus in the zone where the Field of Repulsion of the Sun is greatest. Precisely because of this, because of collision of the biggest in magnitude Fields of Repulsion of the planet and the Sun, the changing of the position of the hemispheres of the planet can't occur by vertical. I.e. the lower hemisphere can't just "go" back and up and the upper - down and forward.

The planet in the process of changing hemispheres implements a "workaround". It rotates so that its own equatorial Field of Repulsion at least extent collides with the equatorial Field of Repulsion of the Sun. I.e. the plane in which the Field of Repulsion of the planet is located at an angle to the plane in which there is the equatorial Field of Repulsion of the Sun. This allows to the planet to save available distance to the Sun. Otherwise, if there was the coincidence of the planes in which the Fields of Repulsion of the planet and of the Sun manifest, the planet would have been severely set from the Sun.

That's the way the planets change the position of their hemispheres relative to the sun - sideways, sideways ...

The time from the summer solstice to the winter for any of the hemispheres is a period of gradual heating of this hemisphere. Accordingly, the time from the winter solstice to summer is a period of gradual cooling. ***The moment of summer solstice*** corresponds to the lowest total temperature of the chemical elements in this hemisphere. ***The moment of winter solstice*** corresponds to the highest total temperature of the composition of chemical elements in the composition of this hemisphere. I.e. ***in the moments of the summer and winter solstices that hemisphere is drawn to the Sun that chilled at this moment most of all***. Amazing, is not it? After all, as our everyday experience says us that everything should be the opposite - in summer it's hot and in winter it's cold. But in this case we are talking not about the temperature of the surface layers of the planet, and about the temperature of the entire thickness of the substance.

But ***the moments of vernal and autumnal equinoxes*** even as correspond to the time when the total temperature of both hemispheres is equal. That is why at this time both hemispheres are at the same distance from the Sun.

And finally we say a few words about the role of heating of planets by solar radiation. Let's do a little thought experiment during which we look at what would have happened if the stars do not emit elementary particles and not heated thereby surrounding them planets. If the Sun did not heat the planets, they would always be turned to the Sun by one side, like the Moon, the Earth satellite always facing the Earth by the same side. Absence of heating firstly would deprive the planets of necessary to rotate around its own axis. Secondly, if there was no heating, there would not be a consistent rotation of the planets to the Sun then one then other hemisphere during the year. Thirdly, if there was no heating of the planets by the Sun, the axis of rotation of the planets would not be bent to the ecliptic plane. Although at the same time all planet would have continued to revolve around the sun (a star). And fourth, the planet would not have increased gradually the distance to the sun.

THE REASON OF PRECESSION OF THE EQUINOXES

The precession of the equinoxes - in other words is the antecedence of the equinoxes.

The precession of the equinoxes (Latin Praecessio aequinoctiorum) – is the historical name for the gradual displacement of points of the vernal and autumnal equinoxes (i.e. the points of intersection of the celestial equator and the ecliptic) towards the annual motion of the Sun. In other words, every year the vernal equinox comes a little earlier than in the previous year.

Both equinoxes are preceded - and the spring and autumn. And this anticipation should be explained by the fact that the planet is gradually heated.

Generally, once we should remember that moment of equinox (any) corresponds to the time when the total temperature of the substance of both hemispheres is equal in magnitude. That is why in the equinoxes both hemispheres are at the same distance from the Sun.

The largest amount of solar radiation is accumulated by the planet's core. This is explained by the existence at any celestial body the centripetal Field of Attraction, because of the action of which elementary particles accumulated on the surface of the chemical elements of the surface layers of the planet flow down, moving towards the center of the planet. Namely because of this flowing of particles down, there occurs a cooling of the surface layers of the planet and a heating of the core. The more heated the planet's core, the less the value of its Field of Attraction. The less is the Field of Attraction, the slower the layers of the planet cool – i.e. the slower particles flow downwards. It turns out that the more the planet's core is heated, the faster there occurs the warming of a hemisphere.

Substance of each hemisphere is external (surface) with respect to the central part of the planet.

An amount of substance in the composition in both hemispheres is the same. And the heating of the core of the planet occurs incrementally – i.e. in every next moment temperature of the substance becomes more and more. That's why when there is a change of the position of poles, and another hemisphere comes to replace a warmed hemisphere, starting to warm up, it warms up a little bit faster. I.e. a little earlier than it was last year, the temperature of this hemisphere compared with the temperature of the second hemisphere. *And this just marks the moment of the equinox. Because of this the days of equinoxes occur all the time a little earlier than before.*

THE MECHANISM OF COOLING OF THE SURFACE LAYERS OF A PLANET

Once the particles fall on the planet and accumulate on the surface of elements in the surface layers either immediately or after some time they begin their way towards the center of the planet. So there occurs a process of cooling the planet's surface (atmosphere, hydrosphere, crust). Particles move down along the line connecting the approximate point of falling to the planet and the center of the planet. The more solar particles with Fields of Repulsion will be accumulated along this line, the greater the substance will be heated, the less solar particles falling on the planet and accumulated in the surface layers will flow down in the process of cooling.

Naturally to the center of any planet the temperature of its substance grows precisely because any spherical celestial body has a centripetal Field of Attraction, which also makes the particles falling on the planet to strive towards the center of the planet.

A GRADUAL INCREASE IN THE ANGLE OF INCLINATION OF THE AXIS OF ROTATION OF PLANETS

In the early life of the planets they had no inclination of the axis. The cause of the inclination is an attraction of one of the poles of the planet by one of the poles of the Sun.

Let's consider as the inclination of axes of the planets appears.

When the substance from which the planets are formed is ejected from the Sun it does not necessarily an ejection occurs in the equatorial plane of the Sun. Even a small deviation from the equatorial plane of the Sun results in that the formed planet to one of the poles of the sun is located closer than to another. And to be more precise, it is only one of the poles of the formed planet is closer to one of the poles of the Sun. For this reason, namely this pole of the planet "feels" a greater gravitation from that pole of the Sun, to whom it is closer.

As a result one of the hemispheres immediately turned toward the Sun. *So an initial inclination of an axis of rotation of a planet appeared.* That hemisphere, which was closer to the Sun, respectively, immediately began to receive more solar radiation. And because of this the given hemisphere from the outset began to be warmed to a greater extent. A greater heating of one of the hemispheres of the planet becomes the reason that the total Gravity Field of this hemisphere decreases. I.e. during the warm-up of the approached to the Sun hemisphere there was a decrease of its tendency to move closer to the pole of the Sun, which gravitation made the planet to be inclined. And the more this hemisphere is heated, the more leveled the gravitation of both poles of the planet, each to "its closest" pole of the Sun. As a result, the heated hemisphere more and more turned away from the sun, and the more chilled began to approach. But note how this change of poles was happened (and happens). Very peculiar.

With the passage of time the angle of inclination of the axis of rotation of planets gradually increased. Let's find out why this is happening.

Generally, when we solve the problem of gravitation (attraction) of any planet to the sun, we should always remember that just two areas of the planet - its magnetic poles - tend to be closer to the Sun. This is explained by the fact that it is the poles are the least heated regions of the planets. Whenever a planet is rotated to the Sun by one of its hemispheres, at the same time gravitation to the Sun of the second hemisphere continues to control this hemisphere. Only this "control" does not allow the planets from the beginning to turn to the Sun by their poles.

Then why, over time, does an angle of inclination of the axis of any planet increase more and more? Yes, because over time there has been increasing the square of the surface of the planet, where there is formed a not disappeared Field of Repulsion. (We recall that the Field of Repulsion of the planet does not exist near the surface of the planet. It manifests outwards in the upper levels of the atmosphere.) A Field of Repulsion creates a Force of Repulsion. A Field of Repulsion of the planet collides with the Field of Repulsion of the Sun. And because of this those areas of the planet, where it has not disappearing Field of Repulsion, tend to move away from the Sun. And so there occurs an inclination. And the farther away from the equator and closer to the poles this area occupied by the not disappearing Field of Repulsion moves, the greater is the angle of inclination of the axis.

We will say a few words about a non-vanishing Field of Repulsion of the planets.

Substance in the equatorial plane of the planet is warmed up to the greatest extent. This is explained by the fact that at the beginning of life on the planet, when it just started to rotate, the axis of rotation is perpendicular to the line drawn through the center of the Sun and the center of the planet. Because of this the equatorial region of any planet from the outset is closest to the center of the Sun. So, the inclination of the axis indicates that at the planet in the equatorial region and adjacent areas a continuously existing Field of

Repulsion is formed. The Field of Repulsion of the planet is formed by particles with Fields of Repulsion accumulated on the surface of the chemical elements of the atmosphere - staying in the composition of the surface layers of the planet. As already mentioned, the planets have an "incline" namely because of arising of this constant Field of Repulsion. Nearest to the center of the sun is always an area of the planet, which is just in the state of the heating. This area accumulates the solar particles, and as a result here there arises a Field of Repulsion, which makes this area to turn away from the Sun.

Why we are talking about this? Yes, because *the parallel of latitude, on which there is an area of the planet closest to the center of the Sun, and just serves as a border between parallels where there is a constant Field of Repulsion (equator and adjacent parallels) and the regions where the Field of Repulsion does not even arise (a pole and polar regions).* On this boundary parallel of latitude a Field of Repulsion appears only after heating by solar radiation. And after some time, due to the rotation of the planet, a cooling this area leads to the disappearance of this repulsion field.

THE EARTH'S GRAVITY DECREASES WITH TIME

All planets including Earth accumulate elementary particles emitted by the Sun. Most of these particles are concentrated in the central part of any planet. The heating of the substance is caused by the accumulation of particles with Fields of Repulsion. Such particles predominate in the composition of the solar radiation. Accumulated particles with Fields of Repulsion reduce the magnitude of the Fields of Attraction of chemical elements on the surface where they accumulate. So, in general, they reduce the magnitude of the Field of Attraction entire planet. From this follows a simple conclusion – a gravity of any planet including Earth with time is reduced more and more. I.e. weight (Force of Attraction) of any body on our planet with every moment is less and less. However, these changes are so small that they are difficult to measure. Perhaps a comparison of weight of the same body, taken at the beginning and at the end of the century will give visible results.

THE SPEED OF ROTATION OF PLANETS - WHAT IS THE REASON

All planets revolve around their own axes. However, each of the planets rotates at its own velocity. Here are the values:

1. Mercury - one revolution around its axis in about 58 Earth days;

2. Venus - turnover for 243 days;

3. Earth - turnover for 24 hours;

4. Mars - turnover in 24 hours 37 minutes;

5. Jupiter - turnover for 9 hours and 55 minutes;

6. Saturn - turnover in 10 hours 40 minutes;

7. Uranium - turnover in 17 hours 14 minutes;

8. Neptune - turnover in 16 hours 03 minutes;

9. Pluto - turnover of 6.38 days.

The speed of rotation of planets is entirely caused only by one thing – by the speed of the heating of their surface layers.

As mentioned earlier, the mechanism of rotation of planets is explained by the occurrence of the Field of Repulsion in the area of a planet, turned at this moment to the Sun. The emerging Field of Repulsion of a planet is resisted by the Field of Repulsion of the Sun and makes this area to move away from the Sun. At the same time the cooler regions of the same hemisphere tends to the Sun. Both of these factors, taken together, make the planet to rotate around its axis.

In each of two hemispheres of the planet there is a parallel of latitude, which is the boundary between the equatorial regions (and near equator), where there exists in the atmosphere a not already vanishing Field of Repulsion, and the Polar Regions, where there is no such field, and there is only a Field of Attraction. Namely on this boundary parallel a Field of Repulsion arises only in the region, which is currently rotated to the Sun. When this area is facing away from the sun, a Field of Repulsion gradually decreases and then disappears, in order to appear again when this area again will turn to the sun.

So, it is the speed of emergence of a non-permanent Field of Repulsion on the boundary parallel determines the speed of rotation of the planet.

And now let's find out on what factors the rate of arising of the Field of Repulsion on the boundary parallel depends. These factors determine the value of the speed of rotation of the planet.

The first factor affecting the speed of rotation of planets is the distance from a planet to the Sun. The distance is not important in itself. The value of the distance to the Sun informs us about the amount of solar particles with Fields of Repulsion reaching a planet. The shorter the distance to the Sun is, than the more solar particles with Fields of Repulsion reach a planet, the more heated the surface layers are and the faster the planet rotates. Conversely, the greater the distance is, than the less number of particles reaches the planet and the slower heating of the surface layers is.

The second factor is the degree of heating of the substance of both boundary parallels separating the regions where there is not disappearing Field of Repulsion from the areas where such Field yet does not exist. Any planet has two such boundary parallels of latitude. The substance, whose degree of heating we are interested, this is a whole thickness of substance that is located under this parallel, up to the center of a planet. Degree of heating of substances means the amount of solar particles with Fields of Repulsion accumulated by chemical elements of the substance. I.e. the more solar particles with Fields of Repulsion are accumulated by the substance of a planet in the area of these parallels, than the faster a not constant Field of Repulsion arises at a planet, and the faster a planet rotates. The greater extent of heating of the bowels of the planet, the less its Field of Attraction is. This means that elementary particles from the Sun reached the planet and accumulated by chemical elements of the surface layers (of the atmosphere) will move down more slowly towards the center of the planet. Therefore, a necessary Field of Repulsion will be formed by these particles faster.

The third factor is the atmospheric composition of planets and its thickness (if it is presented at a planet). The more sparse (the less dense) gases form the planet's atmosphere, the easier this atmosphere can start to produce a Field of Repulsion – i.e. can begin to emit Ether. The explanation is that the smaller the gas density is, the faster these elements form a Field of Repulsion during the accumulation of particles with Fields of Repulsion by chemical elements of gas. In the language of modern physics, the less dense gas is easier to heat. But denser gases are more difficult to heat. This means that for the occurrence of Field of Repulsion at elements forming these gases they must accumulate (absorb) more particles with Field of Repulsion.

As it is known, the most low density gases are included in the atmospheres of the giant planets. Such gases as helium and hydrogen are very easy to heat, and they quickly begin to emit ether – i.e. a Field of Repulsion arises at them very quickly.

Now, if we summarize these three factors and analyze their impact in relation to specific planets of the solar system, we will get something like this.

As you know, the giant planets rotate most rapidly: Jupiter – a turnover for 9 hours and 55 minutes, Saturn - 10 hours 40 minutes, Uranus - 17 hours 14 minutes, Neptune - 16 hours 03 minutes. As you can see Jupiter and Saturn are the fastest. But the distance factor is not on their side. Four planets are closer to the Sun than Jupiter, and five planets are closer than Saturn. Distance from the Sun of other giant planets is more. Nevertheless, even the most remote giant planet - Neptune - rotates faster than any of the terrestrial planets. What's the matter? The reason is a combined influence of two other factors - the degree of heating of the planet and measure of sparseness of its atmosphere.

The farther from the sun there is a planet, the more heated substance is in the area of its boundary parallels. And the giant planets, which are located from the Sun farther than terrestrial planets, are formed from the solar substance earlier and therefore longer feel the effects of solar rays.

And, of course, the atmosphere of the giant planets has a larger percentage of such rarefied gases as helium and hydrogen, and this also contributes to a higher speed of their heating and hence a higher speed of rotation.

Regarding the speed of rotation of such planets of the terrestrial group like Earth and Mars, it is less than that of the giant planets, but much more than that of Mercury and Venus. The Earth revolves around its axis in 24 hours, Mars – in 24 hours 37 minutes. Earth and Mars rotate fast enough due to the greater heating of the substance than that of Mercury and Venus, and also thanks to a sufficiently high degree of sparsity of their atmospheres

The speed of rotation of Mercury is so small - one revolution in 58 Earth days - due to the fact that the substance of Mercury is heated very slightly (less than all other planets), and because Mercury has virtually no atmosphere.

Now about Venus. Its rotational speed is 1 turnover for 243 days. So, the speed of rotation of Venus would have been much more, if it rotated forward and not backwards. This means that at forward rotation Venus would rotate much faster than Mercury. Besides Venus is heated stronger than Mercury and also has a pronounced atmosphere (though dense), while Mercury's atmosphere, we can say no.

Here it should be said about the fact that the speed of rotation of Uranus would be much more if it also rotated in the forward direction, and not the reverse. At the same time Uranus rotates more slowly than the more distant Neptune.

So, slow rotation of Venus and Uranus should explain so.

And now, actually, about why Venus and Uranus rotate more slowly than they would can if their rotation would be direct and not reverse.

For this, we should remember that in the mechanism of rotation of planets once two factors play an equally important role. First, it is an emergence in the heated region of the planets a Field of

Repulsion that makes this area to move away from the sun. And secondly, the tendency of areas of the planet chilled out on the night side to move closer to the Sun.

The Sun's gravity field is an ethereal stream, moving counterclockwise in the direction of the poles and the polar regions of the sun (yes, the Sun also has poles). So, that hemisphere of the planet, it is the side that is in this ethereal stream closer to its source (i.e. to the Sun absorbing the ether) will experience greater attraction from the sun's magnetic poles, as the force of attraction as it is known decreases with distance. The hemisphere of the planets with direct rotation the ***eastern hemisphere*** (moving from the night side to the day) is the closest to the source of the sun's gravity field. While at the planets with the reverse rotation – it is the ***western hemisphere*** (moving from the day side to the night).

Accordingly, the second hemisphere of the planet, which is more remote from the source of the sun's gravity fields, feels far less attraction to the Sun, as the force of attraction decreases with the distance. For planets with direct rotation the more remote hemisphere is western. But for planets with reverse rotation this is the eastern hemisphere.

It is the eastern hemisphere of the planet has a Field of Attraction. And its greatest value in comparison with other areas of the planet, since it is this area was on the night side, and most of all cool. It is the eastern hemisphere through its greatest aspirations to the Sun makes the planet rotated.

In its turn the western hemisphere is characterized by the Field of Repulsion gradually turning into the Field of Attraction (due to the gradual cooling). The western hemisphere also seeks to approach to the Sun, but to a much lesser extent.

And here please note your attention. The planets with direct rotation on the western hemisphere have the area (where the Field of Repulsion disappears and instead it the Field of Attraction appears) is turned away from the Sun to such extent and is separated from the source of its Field of Attraction that for this area the shortest path to the source of the sun's gravity fields is a movement counterclockwise (i.e. continuation of existing movement). The planet does not seek to turn back clockwise.

But the western hemisphere of the planets with reverse rotation is the closest to the source of the sun's gravity field. Consequently the region of the western hemisphere (where the Field of Repulsion due to the cooling of the planet disappears and is replaced by the Field of Attraction) feels a significant Force of Attraction to the Sun. It turns out that the eastern hemisphere of the planets with reverse rotation is located from the source of the sun's gravity field further, which reduces its aspiration to the Sun. And, moreover, and the western hemisphere seeks to the Sun. *As a result, this aspiration to the Sun of the western hemisphere slows down the rotation of the planet, because it prevents the aspiration to the Sun from the side of the eastern hemisphere.*

THE CAUSES OF DISCREPANCIES CONTINENTS AND SEPARATION OF PANGAEA

"Around 1960, an American geologist Harry Hess Hemond suggested that molten rocks of the mantle rose from some cracks, in particular extending along the middle of the Atlantic Ocean. Near the top of these rocks spread to the sides, then cooled off and solidified. The bottom of the ocean thus was expanding and stretching. According to this theory the continents do not drift, and they are just displaced by the rocks of expanding oceans" (Isaac Asimov "Guide to Science").

Substance of the planet is expanding. First, it's because of reduction of mass of the nuclei of elements in the process of radioactive decay. Secondly, due to reduction of mass of elements due to accumulation by the Earth of solar free particles. The surface of the planet - crust – is formed by elements with light nuclei and they came out on the planet's surface during volcanic activity. And these elements with light nuclei in the bowels of the planet during the radioactive decay (decay reason - quality transformation of particles due to the action of forces of attraction). Elements of crust because of the small Force of Attraction of nuclei in the radioactive decay are not involved. Volume of the planet increases due to the growing of the planet's Force of Repulsion and to reduce of the Force of Attraction of elements. The temperature of the elements in the planet's interior is higher than temperature of the elements on the surface. A crust - is dense chemical compounds having a low temperature. Continents - are the primary earth's crust. As a result of increasing the volume of the material interior of the planet, the crust cracks and diverges as the crust on rising yeast dough. This is confirmed by continental drift – in reality they diverge, and not drift. All new layers of substance are postponed under the primary cortex during volcanic activity.

At the bottom of continents the old layers of the cortex are the thinnest, so in these places there are constant zones of volcanic activity.

The cracking crust we know as earthquakes, volcanic activity, and fractures in the crust. And a discrepancy between the individual huge zones of the crust is known as "continental drift". Chemical elements with lighter nuclei arising in the bowels during radioactive decay come out on the surface of celestial bodies during cracking of the crust. Or more precisely, they break the bonds between the elements of the cortex.

The continents do not drift in reality. They drift apart from each other as a result of expansion of the earth globe (such as how the balloon expands during its inflating). The bottom of the oceans is constantly "being created" due to constant output of magma from the depths of the earth. The earth's crust in this diverges like dried crust on rising yeast dough.

DEPENDENCE OF INTENSITY OF SOLAR RADIATION FROM SOLAR LATITUDE

The equatorial plane of the Sun is an area of maximum intensity of solar radiation. This is due to the fact that the solar equator is best "warmed up" by the "energy" of the nucleus of the galaxy. Accordingly, the polar regions of the Sun (and of other stars) warmed up by the "energy" of the galactic nucleus worst.

The Sun (i.e. the central celestial body in our solar system) to the nucleus of the galaxy is the same than the planets are in relation to the star. And the stars are also receivers of radiation of "energy" of the galactic nucleus like planets are the receivers of the radiation from stars.

The nucleus of the galaxy - is the central celestial body of our Galaxy. The Nucleus of Galaxy also has an area of equator, which coincides with the plane of the orbits of stars around the Galactic Nucleus.

Concentration of solar radiation in the solar equator is maximal. This is due to the fact that the "energy" of the Galactic Nucleus, reflected by the sun joins to the radioactive "energy" of the Sun. And at the equator a maximal amount of energy is reflected because this region is most "warmed up".

The same can be said about the concentration of radiation of Galactic Nuclei, Nuclei of Supergalaxy and the Central Celestial Body of the Universe. And also about planets, as they also emit own radioactive "energy" and reflect the energy of stars.

The orbits of stars around the Nucleus of the Galaxy are located approximately in one plane as the planets' orbits are located in one plane.

The crust is formed by: first, non-radioactive elements with the heaviest nuclei of those that erupt on the surface of planets, and secondly, the compounds of these elements with the elements with lighter nuclei. Over the crust there is a shell consisting of elements with the lightest nuclei - the lithosphere and the hydrosphere.

The cause of periodic heating and cooling of each element on the surface of the planet is the rotation of the planet around the star and around its own axis. Physical state of chemical elements on the planet's surface varies periodically. During heating by the rays of a star it becomes sparser. During sunset of a star beyond the horizon - it becomes denser. Physical state of the atmosphere changes more noticeable because of their closer location in relation to the star.

In addition, the entire surface of the planet can be divided into "climate zones". A climate depends on the total length of stay "in the rays of the star" and the average distance of a point on the surface of the planet from the star.

The greater is the angle at which the sun's rays "fall" at any point on the planet's surface, the more "warmed" are the elements at this point and the more rarified is their state of aggregation. Therefore, the closer to the equator, the thinner is the state of aggregation of elements and their compounds. Naturally, in the daytime the aggregate state is sparser, and at night is denser.

And vice versa. The smaller is the angle at which the sun's rays illuminate some point on the surface of the planet, the weaker this area is "warmed up" and the denser the state of aggregation of elements. Therefore, the closer to the poles, the denser is the state of aggregation of elements.

THE REASON FOR EXISTENCE OF RINGS AT THE GIANT PLANETS

At the equator of any planet of the solar system the Force of Attraction (gravity) is less compared with other areas, such as poles. This is due to a stronger "warming up" of the equator. As a result, individual molecules and chemical elements of the atmosphere are not so hard attracted by solid phase of the planet in the equatorial region. The value of atmospheric pressure is reduced closer to the equator. Layer of the atmosphere above the equator is thicker. But also it is thinner as compared with the poles.

Chemical elements and molecules of the atmosphere, located in the uppermost layers of the atmosphere are distanced on such large distance from the center of the planet that they cease to reach the necessary amount of thermal radiation from the planet. And at the same time the giant planets are very distant from the Sun and therefore the atmosphere gets very little sunlight. As a result, the aggregate state of the atmospheric gases becomes solid, and they turn into ice particles of various sizes, which form rings.

In general, the very reason for the existence of such phenomena as the "ring" is a difference in mass of chemical elements forming compounds of the atmosphere. The heavier chemical elements that make up the gas molecules, the closer to the surface of the solid phase of the planet such icy chip will be located in the rings. The easier chemical elements - the further away from the planet it is, in the outer regions of the rings.

THE FINAL GOAL OF THE EVOLUTION OF LIFE ON THE EARTH

In the bowels of massive and simultaneously large celestial bodies in the process of radioactive decay there occurs a birth of chemical elements with lighter nuclei. During this process the Force of Attraction of the chemical elements is reduced, and the Force of Repulsion grows. The distance between the chemical elements increases. This means that the substance in the depths of heavenly bodies is expanding. Light chemical elements are pulled out during volcanic activity from the depths on the surface of a celestial body. The lightest chemical elements such as hydrogen and helium because of the small Force of Attraction and large Force of Repulsion move away from the planet's surface so far that they are not held even in the gas shell of the planet - the atmosphere. We can say they "fly into space". Other heavier chemical elements remain on the surface of the celestial body. The more massive is a celestial body, the more "energy" is released in its depths. "Energy" liberated during radioactive decay moves between the chemical elements on the surface of a celestial body. During the moving between elements the "energy" integrates into the "voids" of chemical elements themselves, heating them and thereby preventing the formation of chemical compounds or destroying of existing ones.

The purpose of the evolution of life on Earth is to create beings who could become carriers of chemical elements the composition of which would be attended by elementary particles of all Plans. Esoteric name for the chemical element is "Soul" and "The Fifth Element". Soul is born from the "marriage of Matter and Spirit". So it happens in reality. "Matter" is heavy elementary particles. "The Spirit" is light elementary particles. Those and others are present in the composition of chemical elements. In chemical elements of any kingdom - mineral, vegetable, animal or human - "Matter" is elementary particles of the lower, most dense layers of the Physical Plan. "Spirit" for the chemical elements of each kingdom is its own. For chemical elements of the mineral kingdom "Spirit" is elementary particles of the higher, ethereal layers of the Physical Plan. For chemical elements of the vegetable kingdom "Spirit" is particles of the Astral Plan. For the Animal Kingdom – particles of the Mental Plan. For Human Kingdom – particles of Buddhic Plan. Chemical elements containing astral particles are included only in the DNA of each cell. Mental and buddhic particles are the part only of the DNA molecules of neurons of different departments of the nervous system of animals and humans. Chemical elements of all other molecules (not DNA) in the composition of bodies of plants, animals and people refer to the mineral kingdom. The bodies of creatures of the Fifth, superhuman kingdom will develop on the basis of human bodies. But the heart of the bodies of superhumans will also be chemical elements of the mineral kingdom.

A COMMENTARY TO THE COSMOLOGICAL HYPOTHESIS

The basis of modern cosmological concepts is the notion that hydrogen and helium - the lightest of the known chemical elements – are served as "building blocks" from which during the process of thermonuclear fusion all heavier chemical elements were created. It is also believed that the source of "energy" emitted by the sun is still the same thermonuclear fusion.

However, this theory of emergence of heavy elements from lighter has a significant disadvantage.

Hydrogen and helium are very light due to the fact that their Force of Attraction is small and Force of Repulsion is large enough. If the Force of Attraction is small, there is no reason that could "make" chemical elements to merge with each other, forming thus the elements with a large number of heavy elementary particles in nuclei - that is, the heavy elements.

But the process of arising of light chemical elements (with less massive nuclei) from heavier (with more massive nuclei) is quite likely.

Once again we focus on the important point. The mass of a chemical element can change both due to changes in the mass of the nucleus, and by changing the quantity and quality of the "energy" in the "voids" of element.

THE REASON OF ABSENCE OF THE ATMOSPHERE ON THE MOON

Our moon like the rest of the bodies in our solar system is an ejection of solar material. It was "captured" by the Earth and began to rotate around it. And it was formed around the same time as Earth. Therefore, it consists initially the same percentage of heavy elements, as in the composition of the earth – that is larger than in Venus and Mercury, but less than in the giant planets. Due to the larger number of chemical elements constituting the earth, the absolute number of heavy elements in its composition initially was more than that in the composition of the moon. Therefore, the processes of radioactive decay in the Earth proceed more intensively than in the bowels of the Moon. As a result, during the radioactive decay the Moon produces less light elements than Earth. This, and the fact that the number of elements of the Moon is less than the number of elements of the Earth and, consequently, the total gravity field is weaker causes that this small amount of gases that formed in the bowels of the moon is not held by the gravitational field.

But on many satellites of Jupiter, for example, on Io, the atmosphere exists. This is explained by the fact that the solar substance from which these satellites formed was rich in heavy elements (like substance of the giant planets themselves).

WHAT IS A COMET?

Belts of comets, which apparently exist beyond the orbit of Pluto, are *the same as the rings of the giant planets - frozen atmosphere of the Sun. The ice blocks of different sizes.*

As meteorites can knock out of rings the individual chunks of ice, and larger celestial bodies (asteroids) can knock out the ice blocks from the belts of comets. A block can be knocked out at the collision in the direction of the Sun. The sun begins to attract it. So there arises a comet.

FORMATION OF PLANETS IN THE SOLAR SYSTEM

There are many hypotheses of origin and formation of planets of the solar system. Let's take a close look on one of them, as it is the closest to reality. It was proposed by the Soviet scientist, Academician Vasily Fesenko. He suggested that the planets may have "solar origin".

Everything is correctly. Planets are formed from material ejected by the star during its heating by radioactive decay of heavy chemical elements in its composition.

Giant planets differ from the terrestrial planets only by their huge thick atmosphere. Under the atmosphere there must exist as solids planets, like the terrestrial planets.

Terrestrial planets are the product of Sun's ejection in a cooled down state as compared with the earlier and hotter state when there was an ejection of substance "for" giant planets. The earlier Sun contained a greater number of heavy elements. Consequently, the substance from which the giant planets formed contained more of heavy elements. Therefore, radioactive decay processes in the bowels of giant planets were more intense. As a result, they have accumulated more "energy"" – i.e. have warmed stronger. Therefore, their chemical elements are in a sparse state of aggregation.

Terrestrial planets formed from material ejected by the Sun in the later periods of its existence. By then the Sun had already lost much of its heavy chemical elements - the source of "energy" - radioactively decayed to lighter elements. Therefore, the substance of the terrestrial planets was less rich in heavy elements compared with the material "for" giant planets. Hence - the lower intensity of the radioactive decay processes in terrestrial planets and less amount of accumulated "energy". And therefore there is a more dense aggregate state.

All planets are heated and cooled simultaneously.

They are heated due to emission of "energy" in the process of radioactive decay of heavy elements. And due to getting of the "energy" emitted by a star into the cosmic space.

Planets are cooled for the same reason that all other celestial objects - due to radiation of "energy" into space.

Radius of all planets increases toward the equator and decreases to the poles. And the gravitational field decreases to the equator and increases to the poles. The reason is the additional heating of chemical elements caused by the "energy" obtained from the Sun. Solar "energy" integrates into the "voids" in elements, leading to additional shielding of nuclei of elements and reducing their mass. Reducing the mass of the body - is increasing its Force of Attraction and decrease of the Force of Repulsion. That's why closer to the equator the planet more "swells" and its gravitational field (Power of Attraction) is less.

Not only the planets, but also all other celestial bodies are heated and cooled simultaneously.

Planets are always younger than stars. Stars are younger than galactic nuclei. Galactic nuclei - than Nuclei of Supergalaxy. Supergalaxy nuclei - than the Central Celestial Body of the Universe. More ancient celestial bodies since the beginning of their existence are heated more than the younger, since in their structure there were still a lot of heavy chemical elements that do not have time to dissolve.

Radioactive decay processes in the Sun proceed more intensively than in any other celestial body in our solar system.

Initially, after separation from the Nucleus of the Galaxy our solar system was one body. Planets did not exist. This body revolved around the Galactic Nucleus on interior galactic orbits. This body - a unified body of the solar system received the emission from the Galactic Nucleus, caused by occurring processes of radioactive decay. These processes take place there and now.

Stars are born due to the volcanic activity in the Nuclei of Galaxies, planets – are caused by the volcanic activity in the bowels of stars. Nuclei of galaxies are born from nuclei of supergalaxies. However, moons orbiting planets are not the product of volcanic activity of planets. They represent the ejections from stars.

The sun is the major primary source of "light" in our solar system. In the depths of the Moon the processes of radioactive decay are still weak and the whole liberated "energy" is absorbed by its own chemical elements. Therefore, the Moon shines by reflected and transmitted solar "light".

THE REASON OF TIDES IS NOT GRAVITATION OF THE MOON, AND THE PRESSURE OF THE HEATED BY THE SUN ATMOSPHERE

Let's talk about the causes of the tides, when every day and every night the water of the world's ocean comes out on the dry land. On the school bench, at physics lessons, students are forced to believe that fluctuations of the water level are explained by attraction of the moon and the sun taken one with another with the centrifugal forces of a rotating planet. Not only pupils and students believe in this hypothesis. The vast majority of scientists trust it.

In the officially adopted explaining of the tidal hypothesis a number of issues cause bewilderment. In particular, occurrence of the tide on the night side of the Earth does not explain by attraction of the moon and the sun, in spite of all the tricks of the quirky human thought. According to the logic on the night side there should not be the tide. But it exists. In addition, the attraction must reject the mass of water in the direction of the attracting celestial body (the Moon or the Sun), and not to force it to comes out on the dry land. As you can see, already two factors do not find their confirmation in the face of accepted theory, put forward by Sir Isaac Newton, and joyfully "eaten" by public opinion.

Sure, attraction from the side of celestial bodies affects the substance of the planet - rejects it. However, not this factor leads to such substantial redistribution of water in the oceans, outward expression of which is a tide of water to the dry land.

The Moon and the Sun of course attract the water of the world ocean. But it should be remembered that the influence of the Sun is complex - its poles attract (characterized by the Field of Attraction), and equatorial regions repel (where the sun manifests outwardly the Field of Repulsion). The Field of Repulsion of the Sun is not so great as to reach our planet. However, the flux of photons of the Sun and of other types of particles falling to Earth transmits to the substance of the planet energy (aka ether). "The transmission of energy" - this is the impact of the Field of Repulsion. We can say that the Sun affects by the Field of Repulsion indirectly through the flux of emitted particles.

And now there is the most important thing in our explanation of the causes of the tides.

The Field of Repulsion (energy) transmitted to the substance of the planet by the flux of particles and heated by it, this is the reason of onset of water on the land, i.e. of the tidal effect. Solar energy heats the material of the atmosphere, is accumulated by chemical elements. The substance expands when heated. ***Expanding gases of the atmosphere, combined with accumulated solar particles emitting ether (with Field of Repulsion) create the total Force of Pressure, which has an impact on water of the planet's ocean - presses on it***. Solar particles emitting ether increase the atmospheric pressure because the pressure of the atmosphere - it is nothing like the pressure of ether emitted by solar photons accumulated by chemical elements of the atmosphere. Ether emitted by solar particles - is the total Field of Repulsion formed in the atmosphere of the planet. This Field of Repulsion affects the hydrosphere. I.e. we can say that *the pressure of ether on the water smoothes it, presses down. The water displaced water by pressure steps on the dry land - so there is a high tide.* The maximum tide should be observed when the atmospheric pressure during the day is the most.

This proposed concept also well explains the fact of occurrence of the tides on the opposite side of the planet. Water, having the property of fluidity, being displaced by the pressure of ether, seeks into areas of the planet where anything will not press on it – i.e. on the opposite side.

The first low tide is in the morning, when the sun moves to the zenith. A number of solar photons falling on the Earth is more and more, and the atmosphere emits more ether, which presses and displaces water. And there occurs the tide of water to the land. At the same time there is a redistribution of water in the oceans - the water rushes in those areas of the world where the pressure on it will be less – i.e. on the night side of the planet. And there is just the time of the night tide. And this tide is caused by the pressure of ether at this moment on the opposite, the day side. I.e. the night tide is caused by the redistribution of water, by extrusion, displacement of its part from the day side to the night.

Pay attention to the following fact. The "moon" hypothesis of Newton offers as a basis for explaining the rise of water (the water shell of the planet) over the usual sea level. I.e. rise of water as a result of its aspiration to the attracting celestial body.

Our hypothesis is based on the exactly opposite mechanism - on the action not of attraction (gravity) and on repulsion (antigravity). Hydrosphere is flattened by the action of pressing on it ether. The displaced water is looking for a place, moves closer to the night side, meets the land on the way – comes out on it. And this is a tide.

A phase of the Moon, of course, plays a role in the height of the tide. As you know, the greatest tide is at full moon and new moon. Moreover at the full moon it is greater than in new moon. This fact is quite easy to explain. At full moon and new moon the Moon is located on the same straight line that passes through the centers of the Earth and the Sun. At full moon it is behind the Earth. At new moon it's in front of it. As a result, in both cases the attraction of the moon enhances the attraction of the Earth. The attraction of what? On what does this factor affect? ***This is important for attraction of photons and other types of elementary particles. The Earth, as well as any other planet or moon, affects by its attraction on particles emitted by the Sun - it supports their inertial motion. There occurs a summing of Forces - Inertia and Attraction. As a result, the total velocity of particles (solar energy) becomes larger. The less number of them is absorbed by the substance filling the space environment of the solar system. This means that more of them reaches the planet and heats the material, in particular the atmosphere.*** In general, you should never forget that our planet owes its heating to the Sun. And this heat does not stop for a moment from the time when the planet was formed from a substance ejected from the solar interior. So that the flow of solar particles on the planet - it's a very important factor.

The more solar particles get into the atmosphere and are accumulated in it, the more the total Field of Repulsion that occurs in the heated atmosphere and displaces the water. In other words, the atmospheric pressure increases.

The complex "Earth-Moon" when the moon is located on the same straight line with the Earth and the Sun attracts with more Force of Attraction than when the moon is at an angle to this line. At full moon and new moon the Force of Attraction with which the Earth acts on solar particles increases. This is thanks to summation with it of the gravitational field of the Moon. The rate of flow of photons increases. The most number of them reaches the planet. And the atmosphere is heated more. Generally we should not equate the heating of the troposphere and the total heating of the atmosphere as a whole. In the lower atmosphere it can rain. However, the total heating of the atmosphere, even in the rain at the full moon and new moon will still be higher than on other days of the month. And therefore the tide will be more.

It turns out that at new moon and full moon the atmospheric pressure is higher than on other days. Moreover it's higher at full moon than at new moon. And all because at new moon the Moon is located in front of the Earth and absorbs the most part of the solar radiation, picking its way to the Earth. However, a large percentage of particles overcomes the attraction of the moon and passes by it successfully, because the Earth's gravity is more than lunar. I.e. even though there is absorption of a significant number of particles, the Moon manages to do its "good work" and attract the solar particles, to support their movement to the Earth.

That's the whole secret of a great tide at full moon and new moon.

CEPHEIDS ARE DOUBLE STARS

We offer you the following hypothesis concerning the nature of **Cepheids - variable stars**. These stars periodically change their brightness.

The star becomes or larger in diameter and colder (dims), or smaller and hotter.

Cepheid (star with variable brightness) - this is a double star. In this system in the center there is located a star that since its formation has in its composition more of chemical elements. Because of this, its mass is greater and the brightness (luminance) is too. A star-satellite rotates around it. It's sparser, colder. It turns around like the Moon around the Earth. This star originally contained a less number of chemical elements. Due to the lower transmutation the luminosity is less too. The substance of such star is better heated by radiation of the Nucleus of Galaxy – it accumulates more photons. This is because the total number of chemical elements is less. And so it's sparser – and due to this diameter of the star is bigger.

Degree of transmutation (transformation) of chemical elements in the star is the more, the more the total number of elements in its composition, and the more the total Field of Attraction of the star. Transmutation (transformation) - is another name for the temperature rise. Scientists still can't explain what the meaning of the phenomenon of temperature is. Proposed by us Law of Transformation logically explains this physical phenomenon.

But we a little digress from the topic of the article. Although this was done deliberately - to recall the meaning of transmutation.

Particles in the chemical elements transmute - change their quality - thanks to receipt of excessive amount of ether.

The more chemical elements are in the composition of celestial body, the higher the degree of transformation of the particles in the composition of its elements. The more elements are in the star, the greater its overall temperature is.

Hence the smaller star brightness with fewer chemicals.

On the contrary - the larger star emits brighter.

That's why periodic eclipse of the less large (and therefore less bright) star by the larger (and more colorful) perceived by the observer from the side as change of brightness.

After all, the observer is far away. And he does not know precisely – is this a single star or a system of two objects. From here and all misunderstandings and divination of astronomers - what does appear before their eyes?

Our proposed explanation of the nature of Cepheids can prove the following fact.

As we know from the history of astronomy, Miss Henrietta Leviticus discovered the connection between the brightness of variable stars and the duration of the period of changes in brightness for stars of Magellanic Cloud. According to her observations, the greater brightness, the longer periods of normal brightness and reduced.

Magellanic Cloud - is another galaxy. Star cluster generated by another Nucleus of Galaxy, and revolving around it, also as our stars are turning around our Nucleus (and formed from its substance).

The brighter the star is, the greater the period of brightness change.

This fact can be explained as follows.

Stars in the composition of Galaxies - are the same as planets in the solar system.

The stars with the largest number of chemical elements are located at the periphery of any Galaxy. Also as giant planets – i.e. the most large – are located on the periphery of the solar system.

These stars are hotter because they are large. If you remember, the more particles in the conglomerate, the more the warming - transmutation. Accumulating radiation (photons) from the Nucleus of Galaxy stars rotate faster. Since the speed of rotation depends on the heating temperature of the atmosphere (if it exists) and of underlying hydrosphere and lithosphere. If the atmosphere does not exist or if it is extremely rarefied – it depends on the heating temperature of the surface layers - as it is the case of Mercury.

The heating temperature of the surface layer depends on two factors - the total temperature of the star (or of the planet) and on the amount of falling radiation. The farther the star is from the Nucleus of Galaxy, the less radiation it receives. Although the number of accumulated photons in it is more and more – namely because it is moving away. But, nevertheless, it is falling energy causes a star (as a planet) to rotate. And at a certain distance from the Nucleus of Galaxy rotation of the star begins to slow down again. As it's in the case of planets. I.e. at first the rotational speed of planets (and stars) gradually increases and then begins to fall. We can see this by the example of the planets in our solar system.

Those stars of Magellanic Cloud that are most of all closer to us look brighter. And they rotate more slowly compared with the stars a little closer to the Nucleus. Again it's the analogy with planets.

Rotating Field of Attraction (ethereal stream) of star, Nucleus or any other celestial body is the cause of rotation of planets around it (stars - in the case of Galaxy). Ethereal stream moves in the direction of the biggest lack of ether – i.e. toward the nearest star. And planets consisting of particles and being immersed in the ethereal field move along with the stream – are attracted. Since the star rotates on its axis – ethereal stream (Field of Attraction) takes the form of a spiral. And planets, respectively, are moving along the same trajectory. More precisely, they are moving in a circle. **But** *they would move in a spiral, and would have fallen onto the star ...* if not heating of them by energy emitted by the star. Namely this heating does not let them fall. **Do you remember the torment of physicists associated with the "need" of falling of electron onto the nucleus of the atom?** In the case of planets the situation is similar. The planets must fall on the Sun. Because the Law of Gravity is relentless. And they do not fall just because the star emits energy (particles). If it did not radiate - it would fall. But the young star and star of "middle-aged" can't but radiate. As for the luminary "aged", then sooner or later it will ebb. But that's another topic that is covered in other articles.

So the planets are held around the star and turn it around due to the gravity. But they do not fall onto the sun thanks to antigravity.

The same can be said of stars in the Galaxy. They do not fall onto the Nucleus for the same reasons - antigravity. This is the Field of Repulsion arising in the surface layers of the star. Emitted ether.

Any celestial body does not fall on its body-grandparent because repelled by ethereal Field of antigravity.

However, we are greatly distracted from the central theme of the article - from explaining the nature of the Cepheids.

Star turns around the nucleus of the galaxy due to the rotation of this Nucleus.

With regard to the mechanism of rotation of the star-satellite, then it is different. It is completely analogous to the rotation of the Moon around the Earth. The star satellite turns around not because of aspirations in the Field of Attraction. The orbit of the satellite can even be located not in the equatorial region of the central star. Like our Moon and its orbit.

The star-satellite is heated by radiation of the Nucleus of Galaxy. Like the moon is heated by solar radiation.

The heating leads to emergence of Field of Repulsion on the heated side of the star-satellite. If the star was single, it just would revolve around its axis. But this star system is double. And there is an attraction from the side of the larger star. I.e. on the one hand the heated side of the star-satellite "wants" to move away from the Galactic Nucleus. And on the other hand the star is held by the attraction of the central star. It is this attraction does not allow the star-satellite just to turn around itself. And it still has one option - to start to move away entirely. I.e. not heated region turns away from the Nucleus and the whole star is moving away. In the case of the Moon the same thing happens.

That's how it happens – the movement of the star-satellite. It just "replaced" the rotation on the periodic distancing and then approaching to the Nucleus of Galaxy. And the "fault" of this is attraction from the central star.

When we look at the night sky, watching the stars within our galaxy, we can't on the basis of only one brightness judge about their distance to the center of the Galaxy. And this is because our sun (and we with it) is also located within the boundaries of the Galaxy. Now, if we look at another galactic system, such as Magellanic

Cloud (or any other), then we can say with confidence that any star in its composition exactly is not part of the structure of our Galaxy.

Only parallax can with sufficient accuracy tell us the location of the star relative to the observer.

Generally the closer the star is, the brighter its light is. However, also the larger it is, the brighter it is. So two stars may be located at different distances from us (and from the center). But their brightness will be the same if one that farther from us – is larger, and the one that is closer - smaller.

The structure of any Galaxy is similar to the structure of any solar system. In the center is a large celestial body-grandparent. Celestial bodies of smaller size generated by it turn around it, moved by gyre of the ethereal stream of attraction. In the case of a Galaxy - the stars revolve around the Nucleus. In the case of a solar system - planets turn around the star, the sun. Pieces of ice of frozen gases of the atmosphere of the giant planets, forming rings in total - are also the prototypes of the planets or stars, moving in a circle.

So everything is built by a single scheme. As below, so above. This is not surprising. The principle structure is repeated. Only scale changes.

Biggest stars born first.

As also from the bowels of any star the largest planets are ejected first.

In the future, these stars (planets) took place on the periphery of the galaxy (the solar system) and are named giants - stars or planets. Luminosity of giant stars is the greatest (about sparse red giants we are not talking). It's namely thanks to their scale – i.e. thanks to the total number of their constituent chemical elements. The reasons for this dependence we discussed earlier in this article.

So we can confidently say - the largest (in general) and brightest (on the whole) stars are located on the periphery of any galaxy. Not all, of course, but the overall percentage of them - yes.

So when Miss Henrietta Leviticus eyed Magellanic Cloud, she could be argued (if she would know our hypothesis) that in general in any galaxy, which is viewed from the side (as you know, we are not talking about our galaxy, which we can't see from outside), the brightest stars are also the nearest to us. And it's all because the biggest stars are on the periphery. The biggest - not thanks to the sparseness and volume. More precisely - not thanks only to this reason. They are the biggest - because from birth they contain many chemical elements.

And less bright stars of an alien galaxy in general have less size (less chemical elements in the composition) and are located further.

Let's summarize.

Cepheid - is a star system consisting of two stars. The star with larger diameter and less bright rotates around the central star, less large and brighter. This star-satellite periodically eclipses the central star. Because of this, from the side the eclipse period is perceived by the observer as dimming. We prove this point of view throughout this article.

The more the main magnitude of brightness of the star, the longer the period of the reduced brightness, and the following period of normal brightness.

The main value of the star's brightness is brightness is the central star. The period of the reduced brightness is a brightness of a star-satellite.

The closer to the periphery of the Galaxy, the brighter stars are located there.

And the slower the rotational speed of all stars is.

And the slower the turning of satellites around the central stars is.

The reasons for this deceleration we have considered in detail in the beginning of this article.

And since the farther from the Nucleus, the less the intensity of radiation reaching the stars, so the heating of the surface layers (responsible for the emergence and increase of Field of Repulsion) is carried out longer than for stars closer to the Nucleus. And because the speed is less with which the star-satellite is moving away from the Nucleus, at the same time turning around the central star.

Have you understood the main idea?

The farther to the periphery of Galaxy, the slower the movement of the star-satellite.

After all, it is an explanation for the fact of relationship between brightness of Cepheids and duration of periods of changes in brightness.

The star-satellite moves slowly because it is slowly heated by radiation of the Galaxy Nucleus.

Thank you for your attention!

The contact information
https://authorcentral.amazon.com/gp/books?ie=UTF8&pn=irid58388648

danina.t@yandex.ru - e-mail

Facebook: https://www.facebook.com/tatiana.danina

Vkontakte: https://vk.com/t.danina

The books of the series "The Teaching of Djwhal Khul – Esoteric Natural Science" - **"The main occult laws and concepts"** - http://www.amazon.com/Main-Occult-Laws-Concepts-ebook/dp/B00GUJJR72

(paperback - http://www.amazon.com/The-Teaching-Djwhal-Khul-concepts/dp/1499625421

"Ethereal mechanics" - http://www.amazon.com/The-Doctrine-Djwhal-Khul-mechanics-ebook/dp/B00I8KSY8Y

(paperback - https://www.createspace.com/4836813)

"New Esoteric Astrology, 1" - http://www.amazon.com/dp/B00JF6RMCY (paperback - https://www.createspace.com/4827294)

"Thermodynamics" - http://www.amazon.com/dp/B00KGHK8EU (paperback - https://www.createspace.com/4838412)

And here is the book of my grandpa, **Michael Novikov, a military paramedic**. You can read his **memories about the Finnish war** http://www.amazon.com/dp/B00JYDITQ6

(paperback - http://www.amazon.com/Memories-Russian-Military-Paramedic-Michael/dp/1499786115)

www.ingramcontent.com/pod-product-compliance
Lightning Source LLC
Chambersburg PA
CBHW072040190526
45165CB00018B/1234